Design & Technology

A review of inspection findings 1993/94

A report from the Office of Her Majesty's Chief Inspector of Schools

London: HMSO

Office for Standards in Education
Alexandra House
29–33 Kingsway
London WC2B 6SE

Telephone 0171-421 6800

Contents

Annex

Introduction

This subject profile provides a review of the findings from the inspection of schools conducted for and by OFSTED during the academic year 1993/4. It continues the publication by OFSTED of subject reports focusing on the quality of provision and standards of achievement in design and technology. It extends information and discussion to include aspects of inspecting design and technology which will be of direct interest to inspectors and may be found relevant to schools. A separate review is available for information technology.

The evaluation of standards, quality of education and provision for design and technology is based on evidence from 31 primary school inspections and 261 secondary school inspections. In nearly all of the primary school inspections the design and technology was inspected by trainee Registered Inspectors, although the inspection was managed by Her Majesty's Inspectors of Schools (HMI). All of the secondary school inspections were carried out by Registered Inspectors and their teams. In addition, evidence from the inspection of design and technology in primary and secondary schools by specialist HMI was used to help in the interpretation of information emerging from the analysis of the main inspection database. This has been supplemented by information from DFE and other surveys to provide data on, for example, capitation and staffing needs.

The findings on primary schools and Post-16 work need to be treated with caution since the sample size is so small.

Those sections of this profile concerned with the standards of inspection of design and technology have drawn on additional evidence; in particular that from the monitoring of inspections and the scrutiny of lesson observation forms, subject evidence forms and design and technology paragraphs in inspection reports, to check for compliance with the requirements of the Framework for the Inspection of Schools.

Subject Report

Main Findings

- Standards achieved by pupils in specific lessons at Key Stages 1 and 2 are generally satisfactory, but overall standards of achievement in primary schools in design and technology are often low. Frequently pupils fail to progress in their development of design and technology capability. This is often linked to teachers' lack of subject knowledge and practical expertise in the range of design and technology activities, poor and cramped accommodation, large class sizes, and insufficient resources to enable pupils to have a sufficiently broad range of experiences (paragraphs 9 to 14, 20, 27 to 32, 50, 60, 64 and 69).

- Standards are better at Key Stages 3 and 4 where these deficiencies are less prevalent, with 80% of lessons seen being satisfactory or better and a broader range of activities being encountered. There are good standards of achievement in Post-16 classes (paragraphs 7 to 17 and 33 to 39).

- At all Key Stages, pupils are enthusiastic and well-motivated in design and technology lessons, particularly when engaged in practical designing and making activities. Their confidence and competence varies with the effectiveness of the teaching, the degree of challenge of the task, and the opportunities given to be creative and to work independently (paragraphs 18, 29).

- The quality of teaching improves as pupils move through the Key Stages because the specialist teachers at Key Stages 3 and 4 are more secure in their subject knowledge. They are able to use a wider range of teaching strategies because of the specialist accommodation, facilities and resources that are available. Where general class teachers in primary schools have received INSET, they are more confident and competent and the quality of organisation, planning and teaching is raised. However, at all Key Stages, too little attention is given to matching the work closely enough to the ability of the pupils and the most able are insufficiently challenged whilst the less able are not given enough support (paragraphs 18 to 32).

- Assessment and recording procedures are developing, particularly at Key Stages 3 and 4. However, the focus tends to be on recording what has been done rather than on the assessment of capability in relation to the National Curriculum levels of attainment. There is considerable uncertainty about what constitutes performance at each of these levels, which leads to inaccuracies and to inconsistencies in some larger departments in secondary schools. Assessment has little effect on teaching and learning: a pupil's attainment in one piece of work rarely has any bearing on subsequent activities (paragraphs 33 to 39).

- Curriculum development has been hampered by the imminence of the revised Order but teachers are becoming more familiar with the body of knowledge that constitutes design and technology as defined by the Programme of Study. In a number of schools across the Key Stages, some aspects of work are omitted from the curriculum because of lack of time, expertise, resources or specialist facilities. Modular or rotational arrangements at Key Stages 3 and 4 are adversely affecting continuity and it is difficult for teachers to plan for progression (paragraphs 42 to 43).

Key issues for schools

Primary schools

- The design and technology subject knowledge and experience of teachers needs to be extended, especially for those teaching Years 5 and 6, and more able pupils.

- Curriculum planning needs to ensure that pupils cover the full range of the NC Programme of Study for design and technology and that the work is progressively more demanding, building on pupils' previous experience.

- Co-ordinators of design and technology need training to give them sufficient subject competence as well as adequate time to support their colleagues and to monitor the work throughout the school.

Secondary schools

- Liaison between those responsible for design and technology work in Key Stages 2 and 3 needs to be improved.

- Many schools need to review their curriculum in Key Stage 3 to ensure that pupils are prepared thoroughly to take advantage of the design and technology work in Key Stage 4 if higher levels are to be achieved.

- More attention needs to be given to matching the work more closely to the ability and past experience of the pupils to ensure that the pupils encounter challenging tasks.

- Now that more time is again being given to practical making tasks, schools need to ensure that all health and safety issues are properly monitored. This involves the maintenance and use of workshop equipment, attention to hygiene in rooms used for work with food, as well as the proper training and health and safety certification of the teachers of design and technology. Many schools need to review group sizes to ensure that pupils can work safely on practical tasks.

All schools

- Strategies need to be developed further to help teachers assess accurately pupils' achievement and progress in design and technology, to standardise these judgements, and to use this information to inform subsequent activities.

- The uncertainty surrounding the review of the Technology Order, now extending over three years, has hindered curriculum development in schools and the much-needed INSET for both primary and secondary teachers. Schools need, as a priority, to decide how they are going to meet these INSET needs.

Standards of achievement

The GCSE and GCE results achieved nationally in design and technology in 1994 are addressed in a section commencing on paragraph 81.

Key Stage 1

1   In Key Stage 1, pupils' standards of achievement in relation to their capabilities were at least satisfactory in 73% of lessons, and good in 20%. This compares unfavourably with other subjects.

2   In this Key Stage pupils were beginning to use materials, mainly card, usually to construct models or items associated with their topic work. They were starting to talk about these products and, when prompted, could identify specific elements and functions of them. However, technical vocabulary was rarely developed adequately.

Key Stage 2

3   Standards were generally satisfactory or better in relation to pupils' capabilities in 63% of the lessons. Good or very good standards were attained in only 7% of lessons. The proportion of lessons with satisfactory or better standards improved as pupils progressed through the Key Stage, being particularly low in Year 3. This compares unfavourably with all other subjects.

4 Pupils achieved high standards where teachers set challenging, well-structured tasks. They were able to develop their design skills; apply them in a range of activities; and select and use appropriate tools and equipment competently when making good-quality products.

5 Standards were low when designing skills were not developed sufficiently to give pupils adequate opportunities to develop their ideas and to translate them into action. At times, the pupils had insufficient skill and knowledge to enable them to choose and handle tools, equipment and materials safely and effectively when making products. In some instances the narrow range of materials and equipment, and the inadequacies or lack of suitable accommodation, limited the opportunities for pupils to experience and develop competence in a range of practical activities.

6 Even where standards in specific lessons were satisfactory, overall standards of achievement in design and technology in primary schools were often low. There was a significant decline in the rate of progress pupils were making from Key Stage 1 to Key Stage 2. Standards were more likely to be satisfactory in Attainment Target Te3, largely because the classes and schools which were under-performing in the wider design and technology activity were often covering craft-based activities reasonably well. A significant number had made little attempt to cover Attainment Targets Te1, 2 and 4, leaving pupils little opportunity to develop design and technology capability.

Key Stage 3

7 Standards of achievement in relation to pupils' capabilities were judged to be satisfactory or better in 80% of lessons, slightly higher in Year 7 than in Years 8 or 9. The proportion of lessons where standards were judged to be good or very good was highest in Year 8 (28%) and lowest in Year 9(25%). These figures are slightly lower than the averages for other subjects.

8 The best work was seen when pupils were making products (Te3). Here their achievement was judged satisfactory or better more often than their performance in Te1, 2 and 4. They responded best to well-structured, practical tasks rather than activities that involved only writing and drawing, especially the average and lower attainers. At

best, pupils were able to observe and research in a relevant manner and they succeeded in producing a wide range of individual designs and products of quality from a range of materials and components. When given the opportunity, some pupils were beginning to develop considerable competence in control technology.

9   Standards were low when the work was unstructured and poorly organised. Here the pupils demonstrated little flair, imagination or decision-making and did not have a sufficiently broad or deep knowledge of materials, tools and processes. Inaccurate work, inappropriate choice and use of materials and weak craft-skills led to the making of poor-quality products.

Key Stage 4

10   For the first time, as part of National Curriculum requirements, technology was taken by all pupils in Year 10, either as a full GCSE course or as part of a combined course. The late arrival of syllabuses hampered planning and depressed achievement.

11   Standards of achievement in relation to pupils' capabilities were satisfactory or higher in 79% of the lessons seen. In Year 11, standards were good or very good in 28% of lessons, compared with 24% in Year 10. These figures are slightly lower than the averages for other subjects, particularly those for pupils of higher ability.

12   The best lessons at Key Stage 4 were characterised by effective project management by both the teachers and the pupils; clear objectives and understanding of the procedures to be used; realistic designing with good application of knowledge and skills; appropriate use of IT such as computer-aided-design and word-processing; attention to detail and safe practice when making high-quality products; and excellent presentation of work.

13   In poor lessons pupils demonstrated little creativity and initiative when designing; research, investigation and evaluation were superficial; craft techniques were weak; and the resulting products were of unacceptable quality. In some of the new Year 10 courses, there was lack of cohesion between some of the course modules with only a limited transfer of skills and knowledge. Therefore, despite the specialist teaching in these courses, fragmented learning depressed attainment.

14   In a minority of schools, pupils were following combined GCSE technology courses in Year 10. These consisted of a short course in technology (which approximates to half of a GCSE) coupled with one of similar length from a complementary but distinct area such as art, music or business studies. Standards achieved in the design and technology short course compared unfavourably with standards achieved in the full courses. The more able pupils were succeeding when their courses were well structured and the teaching was brisk but pupils of average and lower ability were struggling to meet the demands of the course in the small amount of time.

15   Higher standards were also associated with strong senior management support for design and technology with an appropriate administrative and timetabling framework and effective departmental management. Above all, high standards of specialist teaching, based on sound understanding of designing and clear objectives of the content and skills expected, were crucial. Here, efficient use of time, clear deadlines, high expectations by teachers and a brisk pace to lessons, all raised pupils' achievements.

Post-16

16   Standards of achievement in relation to pupils' capabilities were satisfactory or better in 91% of lessons, with a slightly higher proportion in Year 13 than in Year 12. Although the number of students involved in GNVQ courses with design and technology content is rising steadily, the number of lessons inspected over the year was small.

17   In general, the students demonstrated their ability to work independently and to be highly motivated by working with industrial equipment and personnel, applying their knowledge and skills in unfamiliar situations. Where they entered national competitions their levels of interest and standards of achievement increased.

Quality of teaching

18   At all Key Stages, pupils were, in general, attentive, responsive, enthusiastic and well-motivated in the lessons observed. They enjoyed practical designing and making activities in particular, working safely

and collaboratively in pairs and small groups as well as independently. Their level of confidence varied, however, with the effectiveness of the teaching. There is considerable enthusiasm evident amongst the growing number of GCE A-level students.

Key Stage 1

19   In Key Stage 1, teaching in 60% of lessons was judged to be at least satisfactory, with few, 21%, being good or very good. These are much lower figures than for other subjects. Many teachers lacked sufficient subject knowledge to be able to provide a sufficiently broad range of design and technology activities for their pupils and lacked the experience to know when to intervene as pupils designed and made their products. Objectives were often unclear, content and activities were poorly matched to pupils' needs, and the challenge and pace of lessons were low.

Key Stage 2

20   The quality of teaching improved a little in Key Stage 2. It was judged to be at least satisfactory in 63% of lessons and good or very good in 21%. However, this is lower than for other subjects. The amount of unsatisfactory teaching was higher than in Key Stages 3 and 4 but some aspects of the quality of the teaching improved progressively as pupils moved through the Key Stage.

21   In the best lessons, the objectives were made clear to the pupils and the appropriate knowledge and skills were taught well. Designing and practical skills were carefully taught, resulting in pupils being able to achieve good standards especially in Te2 and Te3, but often within a limited range of materials and other resources. Work was sometimes effectively linked to other subjects. The teachers of these lessons had nearly always been involved in recent INSET.

22   Where teaching was less effective, the work was not planned to take sufficient account of the varying levels of ability amongst the pupils. The challenge of the tasks and the resources provided were not closely matched to the individual capabilities of the pupils; too little account was taken of learning experiences in the previous Key Stage, in earlier design and technology lessons and in other subjects. In addition,

more explicit goals needed to be set to provide pupils with a clear understanding of the expectations of the teachers, with demonstrations or exemplar material being used to establish the benchmarks and standards required.

23   Teachers often lacked the necessary subject expertise to teach the range of work implicit within the National Curriculum design and technology Order. In some cases, effective team teaching was being undertaken to develop the required depth of knowledge and practical capability.

Key Stage 3

24   The quality of teaching improved significantly in Key Stage 3. In 81% of lessons teaching was satisfactory or better. Good or very good teaching was evident in 41% of lessons, regardless of the ability of the pupils. These figures are similar to other subjects.

25   Teachers' command of subject knowledge was a positive feature in the good lessons as was their clear understanding of their pupils' abilities and the progress they had made. Here, lessons were well planned, with clear goals and a strong sense of purpose. A wide range of teaching strategies was used to good effect. These included clear exposition, demonstrations, discussions and practical activities which, although tightly structured, permitted creativity and decision-making. Lessons were well paced and sequenced to make the most efficient use of the time available. Effective use was made of a good range of support and stimulus materials, as well as non-teaching staff.

26   There was less effective teaching where teachers had a limited subject knowledge, poor designing skills and lack of enthusiasm for their work. There was little evidence of work differentiated according to pupils' abilities and the teachers had too little knowledge of the progress made by their pupils in relation to National Curriculum design and technology requirements. Particular weaknesses were the lack of awareness of pupils' experiences in Key Stage 2 and a belief by many Year 7 teachers that pupils' records inflated the level of design and technology capability that had been developed. This led to teaching that failed to extend or consolidate the skills and knowledge acquired previously. In some cases, the teaching and learning activities

were targeted well below the levels that pupils had already attained, depressing standards significantly.

27   In many lessons, there was insufficient intervention by the teacher when pupils were designing. Teachers did not judge the timing of their intervention, leaving pupils to flounder. In others, there was too much restrictive and narrowing intervention. Frequently teachers spent too much time dealing with requests for tools, equipment and materials from pupils, as a result of inadequate lesson planning, at the expense of providing the sound guidance required to enhance progress.

Key Stage 4

28   The quality of teaching in Key Stage 4 is much the same as in Key Stage 3. Teaching was at least satisfactory in 80% of lessons and good or very good in 39%. These figures are comparable with other subjects. The quality of teaching is better in Year 11 than in Year 10. The problems associated with the late arrival of syllabuses for the new National Curriculum courses for Year 10 clearly impeded teachers' planning.

29   Better lessons at Key Stage 4 were characterised by pupils being given opportunities to think imaginatively and independently and to take responsibility for their own learning with good use being made of resources in the school and the community. Less satisfactory learning was evident when pupils' enthusiasm was adversely affected by heavily teacher-directed and narrow project work. Short, fragmented, modular arrangements in some of the new courses in Year 10 meant that pupils were not able to become fully engaged in projects, and could not develop depth of understanding or capability. In addition, for the first time in most schools, all pupils were being required to study design and technology to age 16, and some of them resented this.

30   The features of good and poor teaching in Key Stage 4 were very similar to those observed in Key Stage 3. A significant factor in poor lessons at Key Stage 4 was the lack of structure and sequenced coverage of knowledge and skills in long projects. There was fragmented learning in short projects where modular arrangements operated, because of inadequate planning by the teachers for the reasons indicated above. In some combined short courses teachers taught the

technology short course without reference to the interests of their pupils in, or the demands of, the combining subject. This led to lack of coherence and lower standards of achievement.

Post-16

31   The quality of teaching in the sixth form is good, with 89% of lessons judged to be satisfactory or better and 46% good or very good.

32   Lessons were well planned and enthusiastically and effectively taught. Tutorials were positive features and students were encouraged to seek additional support and guidance from outside school, particularly from industry. Lessons provided considerable scope for students to think, investigate and experiment.

Assessment, recording and reporting

33   The procedures used to assess and record pupils' attainment in design and technology vary considerably in their coverage and reliability but schools are generally meeting the statutory requirements for National Curriculum assessment and recording at Key Stages 2, 3 and 4.

34   There are, however, particular weaknesses at Key Stage 1 because some teachers are uncertain how to assess pupils against the Statements of Attainment, especially in Te1 and Te4. Sheets used for assessment and recording in primary schools tend to be based on tick-sheets which are cumbersome; they focus more on recording contexts and outcomes than assessment of pupils' attainment and progress.

35   Schools' own assessment and recording systems vary in quality and complexity and some do not assist teachers to determine levels of attainment and progress accurately. Others are the product of considerable change and modification which has helped to streamline the assessment process and provide teachers with an accurate assessment of pupils' levels of attainment in the National Curriculum.

36   In good departments key assessment criteria are identified; they are integral to the planning of schemes of work; and they are integrated successfully into teaching and learning in lessons. In a sizeable minority of others, procedures are not sufficiently systematic or detailed and assessment criteria have little or no influence on the planning and

delivery of the design and technology curriculum. Rotational course arrangements also prevent teachers knowing enough about their pupils' achievement and progress in earlier work. There is little monitoring of assessment procedures at any Key Stage to ensure that teachers comply with school and departmental policies or to identify strengths and weaknesses in teaching and learning. Teacher assessment is rarely used to measure the progress that pupils are making during the Key Stages.

37   A considerable number of teachers have difficulty in developing an understanding of what constitutes performance at each National Curriculum level of attainment in the Attainment Targets and this causes further inconsistencies. There has been only limited development of standardisation and moderation procedures to facilitate the development of this understanding. Those moderation networks between primary and secondary schools that do exist are rarely used to develop an understanding of pupils' design and technology capability. Similarly, neighbouring secondary schools have seldom exploited the opportunity to share perceptions of attainment.

38   Oral feedback to pupils in design and technology lessons is often constructive, helping them to make good progress, but there is little formative assessment of pupils' design and written work as tasks or projects proceed. Marking is sometimes cursory and does not help pupils to judge their progress or to see how to improve their work.

39   Self-assessment is used in many schools as a means of involving pupils in an evaluation of their own performance. However, it is rarely well-developed and is frequently superficial, often being confused with evaluation of the products (Te4) that pupils have produced.

Curriculum content

40   Schools are taking increasing account of the Programmes of Study defined by the National Curriculum Order in their planning. Mapping of the coverage of the Programme of Study is done to varying effect but, in general, omissions of the past are being identified and attempts made to rectify deficiencies and to deliver a comprehensive design and technology curriculum. However, some aspects are being omitted because of lack of staff expertise or inadequate resources,

especially in some single-sex schools and primary schools. On occasions, work is limited by the large group-sizes.

41   In Key Stages 1 and 2, even in schools where standards are generally good, there were gaps, often large, in the coverage of the Programme of Study. Very few schools have schemes of work that sufficiently guide teachers in the development of coherent and progressive design and technology activities. In primary schools topics or projects are used frequently to deliver design and technology, integrated with elements of other subjects. However, the demands of the topics often distort the work and set limits on the standards that pupils can achieve in design and technology.

42   In Key Stage 3, the curriculum tends to be delivered in one of two ways. It may consist of modules to give pupils experience of working with a number of different materials with specialist teachers, or it may be based on projects or themes. Modules are sometimes isolated areas of study that are focusing increasingly on manipulative skills in an attempt to raise standards of making. Where these skills are developed in isolation from designing, overall design and technology capability is reduced. In some schools, however, there is an appropriate balance between teaching skills and encouraging creative thinking.

43   When there is no clear identification of the knowledge and skills likely to be required this can lead to low standards, poor coverage of the required subject content, and slow pace with too much time spent on written and graphical work, often of retrospective design-sheets, at the expense of making activities. In these instances, it is difficult for teachers to plan for progression in the skills associated with research, investigation, drawing, design, making and evaluation.

44   At Key Stage 4 a few schools are failing to meet the statutory requirements for design and technology. In other schools about 10% of the pupils are following GCSE courses in technology that cover Te1–5, ie including IT. Most pupils (73%), are following design and technology courses which focus on Te1–4, with a further 15% combining design and technology with a related subject. Only 2% of pupils have chosen to take a GCSE in technology (Te1–5) with a related subject.

45   Planning for the new GCSE courses was hampered by the late arrival of the syllabuses but schools have generally succeeded in implementing them with reasonable success, with an increasing awareness of the need for greater continuity between Key Stages 3 and 4 and between the elements of extended and combined courses.

46   Inadequate links between the various components of GCSE courses sometimes limited pupils' progress: for example, between the core studies and associated options of extended courses and between the elements of combined courses. Whilst some pupils valued the opportunity to engage in a wide range of experiences, others were stridently resentful of being expected to take a course against their interest. Time to enable teachers to meet, plan and review in order to rectify deficiencies and to plan improvements in these new courses is minimal but much needed.

47   Most schools make strenuous efforts to ensure that all pupils have equality of access to the design and technology curriculum but some are less successful than others in doing so. For example, some projects are too male-orientated and too little account is taken of the cultural background of the pupils. There is a traditional gender bias in most schools within the optional elements of the Key Stage 4 courses and this trend continues post-16, although this bias is steadily reducing.

Provision for pupils with special educational needs

48   Policy statements and classroom support focus on the lower-attaining pupils with little reference to the most able pupils. Differentiation is primarily by outcome, and teachers would do well to consider more differentiated tasks or teaching materials. Frequently the high-attainers are insufficiently challenged and lower-attainers are not supported enough. Most pupils with special educational needs cope well with the making aspects of the work and teachers are effective in providing appropriate guidance. There is less effective support in the initial design work and in ensuring that written information is accessible to all pupils who have reading or language difficulties.

Management and administration

49 Many subject co-ordinators in primary schools have a wide range of additional responsibilities. Few have any non-teaching time to work alongside their colleagues to support them and to share knowledge and skills, often gained through INSET. The GEST 20-day courses for co-ordinators have been particularly effective.

50 Most departments in secondary schools are managed satisfactorily, usually organised by subject co-ordinators who are normally effective in ensuring that the design and technology curriculum complies with statutory requirements, and in managing finances efficiently. Less effective management is characterised by lack of clear job-descriptions which detail roles and responsibilities; by vague performance indicators which make meaningful evaluation difficult; by poor departmental documentation; and by the absence of formal monitoring procedures, leading to low standards of achievement.

Resources and their management

Teaching and non-teaching staff

51 In primary schools most teaching is carried out by generalist class teachers. A few of these are confident and competent within design and technology, almost always as a result of their having received recent INSET or advisory teacher support. Most teachers, however, have not yet grasped the nature of design and technology and lack sufficient subject knowledge to extend the older and more able pupils. In a small number of schools semi-specialist teaching is provided for pupils at Key Stage 2. This is usually well informed and effective.

52 Most secondary schools had enough design and technology teachers this year, although providing for compulsory design and technology at Key Stage 4 has stretched staffing, in some beyond their limits. Most teaching is by specialists with adequate qualifications and experience in one or two aspects of National Curriculum design and technology.

53   However, more than half of the teachers feel inadequately trained to teach electronics, computer control, and CADCAM, although expected to do so. Over two-thirds of teachers said that they had needed to teach themselves some aspects of design and technology in order to deliver the National Curriculum. A significant number do not have an adequate grasp of the overall nature of the subject and, in particular, of how to engage pupils profitably in research, design and the newer technologies. Training needs are acute and only half of design and technology teachers said that they have attended any INSET in the past four years designed to help them teach National Curriculum design and technology.

54   Technicians are usually effective particularly where their time is purposefully managed by heads of departments. Some departments, however, have too little or, in 6% of secondary school and 62% of middle schools, no technician support, leaving teachers to do the essential preparation and maintenance work at the expense of teaching and curriculum development.

Resources for learning

55   The quantity and quality of materials and equipment are severely restricted in many primary schools. At Key Stage 1 schools often rely too much on reclaimed household items, failing to do justice to the whole range of design and technology activities. A few have acquired virtually no resources for current needs. Deficiencies vary widely but include tools, construction materials and equipment for work with electrical circuits and mechanisms. The average capitation is about £2.50 per pupil, ranging from less than 10p to over £10. Insufficient capitation for consumable items and restricted funding for new equipment, associated with the lack of teachers' knowledge in the subject, lie at the root of poor provision for the subject in many schools.

56   Where schools have acquired and made effective use of appropriate resources they have often organised, labelled and stored them to give pupils ready access, helping them to work effectively within design and technology.

57   In secondary schools most design and technology departments make efficient use of their resources. Some have been able to refurbish rooms with efficient, modern equipment. Such provision is often linked to satisfactory or better standards of designing and making and has increased the rigour of pupils' work. However, many departments or sections of departments are not able to resource the breadth and depth of activities needed for the National Curriculum. Such schools are insufficiently equipped for one or more of: research; design drawing; computer-aided design; computer-aided manufacture; electronics; control; food technology and plastics. The gap between the best and worst provision of equipment in schools is widening. Capitation for design and technology averages just over £5 per pupil. However, there are wide variations, from less than £1 to more than £18. Pupils often purchase materials and components for their own projects, extending the range of products which schools are able to finance from capitation. Examples were seen of expensive new equipment lying idle because there was not enough money to buy the necessary consumable materials.

58   Given the constraints on funding, many schools do not consider books to be a high priority for spending within design and technology. Thus, many books are outdated for current requirements, with central library stock being particularly dated and under-used, restricting the range and quality of researching and designing which pupils undertake. However, some schools are developing multi-media design and technology resource areas which provide stimulating bases for pupils to research and develop aspects of their design projects.

59   Whilst a growing number of schools have good equipment for computer-aided design and computer-aided manufacture, some producing superb work, many have little more than basic IT systems used mainly for word-processing or simple graphics work. Computer-control work is partially limited by the lack of appropriate equipment.

Accommodation

Primary schools

60 Most design and technology work is done in classrooms, adequately at Key Stage 1; but in many Key Stage 2 instances, classrooms are too cramped for the practical work necessary for National Curriculum design and technology. Safe and secure storage of materials, equipment and pupils' work is a major difficulty in many schools.

Secondary schools

61 Many middle schools have insufficient specialist rooms and some single-sex secondary schools, mainly boys' schools with no facilities for work with food or textiles, are unable to cover the range of work envisaged in the National Curriculum for design and technology. A high proportion of girls' schools have now provided facilities for work in control and in resistant materials. There are sufficient design and technology rooms in most secondary schools for current needs and they are usually and sensibly confined to specialist use. However, most need enhanced provision in some way in order to meet the new demands of the subject. With the introduction of statutory design and technology in Key Stage 4 there is increasing pressure on specialist accommodation, particularly storage. This year, some schools have been unable to fulfil their statutory requirements in Year 10.

62 The average size of groups continues to increase in design and technology lessons in secondary schools, with a marked rise in the number of schools where all groups in Key Stage 3 contain more than 20 pupils. Groups of over 26 are not uncommon and 35% are more than 20, usually in rooms designed for 20 pupils. In these situations teachers find it difficult to resource lessons adequately, or provide opportunities for sufficient practical work. Sometimes, they are rightly worried for pupils' safety.

Inspection issues

Inspection development

63   Inspections carried out under Section 9 of the Education (Schools) Act 1992 began in September 1993. Inspection teams have made a good start in meeting the requirements of the Framework for the Inspection of Schools; they have become more confident as the year has progressed and some early uncertainties have been resolved in many cases. This part of the subject profile draws together some of the key issues for further improving the quality and standard of inspection. Many issues are similar from one subject to another; where there are subject-specific matters these are indicated.

64   Some examples of inspection writing are included. They are not intended to be viewed as models or templates but illustrate how Framework and inspection documentation requirements can be reasonably met.

Evidence gathering

65   Inspectors generally sample a good range of technology work of different year groups, abilities and Key Stages across the compulsory years of education. They usually achieve a good balance. It is important, though, that evidence covering the full range of the school's activities in technology is drawn from lessons or other sources, including work with construction materials, food, textiles, electronics or other control activity. Where a school has a sixth form, post-16 work needs to be fairly represented in the sampling.

66   In reaching their judgements, inspectors use evidence from a good range of sources. Whatever the sources of evidence, it is important that in inspection writing clear reference is made to them to support judgements. The Supplementary Evidence Form provides a means of documenting evidence and judgements from sources other than lessons and could be more widely used.

Lesson Observation Forms

67   Overall, Lesson Observation Forms are completed conscientiously, with attention to the relevant evaluation criteria. Inspectors could usefully check that subject detail and characteristics are incorporated wherever possible.

68   In relation to the **content** of lessons, the majority of inspectors adequately indicate the topic of lessons usually with reference to the National Curriculum Order. However, further details of the lesson activities would be helpful in setting the context, for example by specifying the main activity, the materials being used, the context for the work or the stage in the project. An example of a 'Content' section follows.

Year 8, mixed ability (textiles lesson)

Working individually, pupils were starting to make a tapestry to extend their understanding of fabric construction. They had previously sketched several possible designs for their product and had developed one of them in detail. Most pupils were at the stage of investigating the effects produced when stitching was done using woollen yarns of varying thicknesses. Te2 and 3 – aimed at levels 3/4.

69   Inspectors draw on their professional knowledge and experience to make overall judgements about the **achievements** of pupils. Responding to the Framework requirements to assess pupils' achievements in relation to national norms and taking account of pupils' abilities has not proved easy. Revised requirements and guidance published in June 1994 should help inspectors in making these distinct judgements. To support judgements it is important that inspectors clearly identify and record what pupils know, understand and can do and set achievements in the context of National Curriculum Statements of Attainment. A pair of examples of 'Achievement' sections from a Lesson Observation Form follows.

Year 7, mixed ability (resistant materials lesson)

Achievement (age referenced): Te1 – pupils were able to ask questions to help to identify needs (level 2) but were not using this information to support their designing and making activities. Te2 – expressing ideas/drawing – nature of the task limited achievement to level 1/2. Te3 –7 pupils were using simple hand tools with varying standards of accuracy and competence (level 3). Te4 – pupils were able to make simple judgements about what they liked and disliked about their design but could not easily evaluate how well the original intention has been met (level 2). Grade 4

Achievement (taking account of pupils' abilities): *Despite a range of abilities there was little difference in the quality of outcomes. One or two more able pupils were setting their own targets but they were not sufficiently challenged by these or by the teacher's expectations. Insufficient support for least able adversely affected their understanding and the quality of the outcomes. Grade 4*

70   Clear evidence of pupils' attitudes to learning and their behaviour in lessons is usually given, and this is reflected in the grade given for **quality of learning.** Greater prominence should be given to other attributes of learning, particularly pupils' progress in lessons and the development of the technology skills included in section 6.5(i) of Part 4 of the *Handbook for the Inspection of Schools.*

Year 9

Pupils applied themselves to the task with enthusiasm, seeking help as appropriate. Most appeared to be enjoying the task and could see the purpose in the context of the last lesson (other techniques of decorating fabric). They persevered and improved their original suggestions. All made progress in cutting their stencil. Grade 2

71   Inspectors usually cite relevant evidence when judging the **quality of teaching,** and evaluation is based on the criteria in the Framework. They need to check that the full range of criteria is used, including teachers' command of the subject. In the following examples a range of attributes of teaching are referred to.

Year 8

Good planning and preparation of materials. Appropriate recap of work so far. Expectations clearly explained to the whole group. Also good small group teaching with skilful questioning where pupils did not fully understand. Pupils were well supported and encouraged to work at a suitable pace. Good relationships. Thorough teaching of designing skills and obvious command of the subject demonstrated. Pupils were encouraged to use sketching as an effective means of developing ideas. Homework set with clear explanations of what was required. Grade 2

72   The Lesson Observation Form could be more widely used to indicate contributions made by the subject lesson to key skills and to learning in other areas of the curriculum. It also provides opportunity to signal the impact of contributory factors on achievements and the quality of learning which can be drawn on when compiling the Subject Evidence Form.

Guidance to inspectors on new inspection requirements

73   The Lesson Observation Form/Subject Evidence Form proforma require the use of standard codes to identify the Technology course being taught (not the content of the lesson). The subject/activity box has two sections. For all Key Stages the first box records the main National Curriculum subject towards which the lesson contributes – **this year, for all Technology, whether IT or DT, this must be 'TE'** and the **second box** will normally be DT. In Key Stage 4, if the course addresses TE1–5 this second box will be left blank. This gives no opportunity to indicate the nature of the course in Key Stage 4, if it is a combined or extended course for example, and so it would be helpful if this information could be included at the beginning of the 'Lesson Content' section of the Lesson Observation Form. Where pupils are following a non-National Curriculum course, Craft, Design and Technology or Home Economics in Year 10 for example, the appropriate code should appear in **box 2**.

Subject Evidence Forms

74   Subject Evidence Forms are usually fully completed, very often thoroughly and thoughtfully. In most cases, a wide range of evidence appears to have been used. Inspectors need to check that reference to this range of evidence is sufficiently explicit in the relevant sections of the Form and to ensure that the emphasis is towards evaluation rather than description.

75   Particular attention is given to aspects of standards of achievement and the quality of learning and teaching although, as in Lesson Observation Forms, when considering the quality of learning more emphasis is placed on pupils' attitudes and behaviour than on their skills as learners. When commenting on examination results as part of their evaluation of standards of achievement, inspectors should ensure that the evidence includes the basis for any comparisons with national data.

76   Extracts from sections on standards of achievement and the quality of learning and teaching from Subject Evidence Forms follow. They provide adequate evidence on which to base the subject paragraph for the report.

Standards of Achievement

In Key Stage 2, standards of achievement were below national expectations in 80% of lessons. In Key Stage 3, standards matched expectations in only 14% of lessons, the remainder being below average. In relation to pupils' capabilities, standards in Key Stage 2 were satisfactory overall (80% of lessons) but in only 50% of lessons in Key Stage 3 were standards appropriate. In Te2, pupils were mostly achieving level 2; there was some evidence of level 3 at Key Stage 3, with annotation of drawings and simple reasoning, and limited examples of level 4. Underachievement in designing was linked to lack of designing skills (modelling, sketching, formal drawing techniques). Pupils rarely used their designs to assist making. There was some evidence of pupils modifying their ideas but these were not recorded or valued as part of the designing process.

Quality of Learning

In Key Stage 3 pupils used an increasing range of techniques, processes and resources with confidence, showing creativity in designing products to meet particular needs. They had the capability to apply knowledge in problem solving situations; could explain their ideas through graphic communication; and realised their projects in a range of materials. Sometimes they failed to recognise accepted craft practice but this was often compensated for by producing good results through the use of ingenious methods. Pupils were able to work independently and as part of a team. They had learned how to work safely and to organise their time in practical situations. Often pupils learned from each other and this was particularly evident in some of the computer-aided manufacture activities where excellent results in textiles had been produced.

Quality of Teaching

Assignments were carefully prepared, and individual lessons were well planned. Pupils had clear design briefs and the tasks set were interesting and relevant. In KS2 pupils were provided with helpful booklets for recording designs, plans, modifications and evaluations. Important features of design and technology, such as costing, and working with limited resources were appropriately covered. The most significant weakness was that although most process skills were taught quite effectively, the teaching of the 'making' skills and principles needed to underpin the work was insufficiently systematic, although there were some examples of good practice.

77   When considering features such as the resources for learning, management and, indeed, the quality of teaching, the emphasis of evaluation should be towards the effects on the standards achieved and the quality of learning. The following extracts from the 'Contributory factors' sections of the Subject Evidence Form include clear judgements about features of provision and some indications of their effects.

Resources for learning

The level of capitation was over one-third lower this year than last, needing very careful financial management to ensure that the planned courses could be delivered. In the areas of construction and textiles in Year 9 there was some evidence that shortage of materials constrained the projects carried out. There were no additional texts to support the new GCSE construction core in Year 10. The department was quite well resourced with IT hardware and software but the area of control, including electronics, and computer-aided manufacture in both construction and textiles could not be developed without an investment of resources. The range and quality of work was currently limited by these shortages. Short-term resourcing difficulties had been identified and managed but there was a need for longer-term planning which both linked resourcing needs to development planning and included consideration of maintenance as well as replacement and development.

Provision for pupils with special educational needs

The grouping of pupils into three broad ability bands for the subject provided some opportunities to match work to pupils' needs. There were some good examples of adjustments to teaching approaches and individual support of pupils, but differentiation by task and, in particular, by resource was insufficiently exploited, either to make work more accessible for the least able or to challenge the more able. In classes of lower ability pupils the favourable teacher/pupil ratio was used to good effect and pupils were generally well-behaved and worked on task. Teachers supported and encouraged these pupils to produce their best work. Generally the pupils were motivated and achieved satisfactory to good standards in relation to their abilities.

Judgement Recording Statements

78 The Judgement Recording Statements are usually fully completed. Inspectors need to ensure that all available evidence is considered in arriving at judgements for inclusion in the proforma. The purpose and use of Judgement Recording Statements are outlined in Appendix C of Part 3 of the *Handbook for the Inspection of Schools*.

Subject sections in inspection reports

79   Most technology subject sections in inspection reports meet the Framework requirements and those seen were well matched with the evidence in the Subject Evidence Forms. They give appropriate emphasis to standards of achievement and the quality of learning and teaching. Inspectors need to ensure that overall judgements are clear and succinct and draw on all the evidence available, and that factors which impact on standards of achievement and quality of learning are clearly identified.

80   The following extracts from four reports illustrate writing about standards of achievement, quality of learning and teaching and a contributory factor. The characteristics of technology are evident.

School A

At Key Stages 3 and 4 all or almost all pupils achieve at least the national expectation in design and technology and many beyond it, achieving high levels for their abilities. At Key Stage 3 pupils show skill in designing and making artefacts using a wide range of materials and processes. Levels of skill in graphic communication are particularly good. Pupils are effective in undertaking research but too often develop their ideas solely through drawing rather than by practical experimentation and evaluation. They produce good quality products, reflecting clear instruction in craft skills and a sensitivity to aesthetic values. There is a growing awareness of the effects of technology on society. Results in the subjects that contribute to technology are better than the national average for maintained schools at both GCSE and A level. However, the A-level design and technology examination results in 1993 were poor ....

School B

Pupils learn to work safely and with confidence when using tools, equipment and machinery. Pupils are highly motivated and they master effectively the knowledge and the skills of designing and making through individual projects and group activities. These require them to make their own decisions and manage the sequence of their work skilfully. Pupils enjoy these challenging tasks and apply what they have learned in a disciplined and creative way.

A-level students become effective independent learners who pursue their projects with vigour and expertise. They are always able to justify their own decisions and to discuss their work with maturity ....

School C

The quality of teaching is satisfactory and often good across all Key Stages. Where teaching is good, the well-qualified teaching staff are sensitive to the needs of individual pupils and take care to explain technical terminology and to check understanding. The majority of lessons are well planned, matched to pupils' abilities and proceed at a suitable pace. Health and safety procedures and practices are sound. However, pupils' achievements in one aspect of the subject are not used to plan further assignments in other aspects. There is a need to implement and monitor an agreed faculty policy for record keeping and marking ....

School D

The faculty handbook contains clear guidance for teachers, but there is no substantial departmental development plan which identifies longer-term aims and staff development needs. There is no overall scheme of work indicating how the various contributions are co-ordinated. Subject schemes of work could usefully be reviewed to clarify the links between what is to be taught and what is to be assessed. The newly refurbished accommodation is bright and uncluttered and provides an attractive environment that is conducive to effective working for staff and pupils ....

81   In writing to the Framework requirements, inspectors need to check that a comment is included on compliance with statutory requirements and that key issues for action in the subject are clearly given. These are helpful to schools in their action planning.

The interpretation of subject performance data

GCSE

82 The interpretation of the examination results at GCSE in design and technology-related subjects presents peculiar difficulties because of: the variety of the changes from existing pre-National Curriculum courses to those that satisfy National Curriculum design and technology criteria; the variety of courses available; and the changes that are steadily occurring in the proportion of pupils taking design and technology-related subjects to GCSE. In 1994 the percentage of the age cohort entered for the main two options was: technology 35% and home economics 19% (for current purposes OFSTED has classified technology as including all the craft, design and technology-related courses but not home economics or information technology).

83 For all pupils in all maintained schools the average points score per entry for technology was 4.05, and for home economics was 4.07. This compares with 4.59 for English, 4.09 for mathematics and 4.22 for science. There are significant differences in the points score according to the type of school and the gender of the pupils. In technology girls, with 23.2% of the entry, do significantly better than boys with an average points score of 4.53 compared with 3.91 for boys. For technology the average score in comprehensive schools is 4.03, for modern schools it is a little lower at 3.60 and in selective schools it is much higher at 5.38. The figures for home economics are 4.04 for comprehensive schools, 3.72 for modern schools and 5.86 for selective schools. Girls do much better than boys with a score of 4.18 compared with 3.37 for boys.

84 Results attained in maintained schools in technology courses at GCSE rose slightly in 1994. Although a smaller proportion of candidates gained A*–G grades, 91.6%, compared with 92% in 1993, the rise in grades A*–C was from 36.8% to 37.4%.

85 In home economics at GCSE, results were similar to technology, with grades A*–C rising from 37.2% to 37.9%, and with a lower proportion of candidates, 94.1%, achieving grades A*–G in 1994 compared with 94.5% in 1993.

86   When comparing a school's GCSE results in design and technology-related subjects with the national statistics inspectors should:

- compare the proportion of the cohort entered for GCSE with national figures;

- consider the option arrangements operating in the school when pupils made their GCSE subject choices;

- recognise that national figures are for candidates entered for courses covered by GCSE subject criteria and that 'other' design and technology-related courses may have different national averages;

- exercise caution when making year-on-year comparisons when there has been a syllabus change, noting the slight fall in average grades nationally.

GCE A/AS

87   The GCE A-level entries in design and technology continue to increase in both the maintained and independent sectors, although numbers are still comparatively small. Entries for home economics are much smaller and are falling.

88   At GCE A-level, results attained in design and technology examintions were similar to those achieved in 1993 with 81.9% of the students achieving grades A–E and 22.1% achieving grades A–B. Again, girls, with 19.4% of the entry, do better than boys. There was an improvement in the home economics A-level results at the higher grades compared with the previous year with 80.4% of the students gaining grades A–E and 23.2% grades A–B. Overall, the results for the higher grades are lower than the all-subject averages (23.2% compared with 29.3%) but are marginally better for A–E grades.

89   GCE AS entries in design and technology remain at a very low level nationally and any comparisons need to be made with great caution. However, results were considerably better in 1994 with 21.3% of entries graded A–B and 78.2% A–E, compared with 14% and 68% respectively in 1993. In GCE AS in home economics 15.3% gained A–B and 58.3% gained A–E, compared with 14.1% and 51.3%.

Update on developments in Design and Technology

Dearing Review of the National Curriculum

90   The Dearing Report invited schools to consider applying, under section 16 of the Education Reform Act, to work to the draft proposals for design and technology for Key Stages 1 to 3 from September 1994. Approximately 4,000 schools, 17% of primary schools and 25% of secondary schools, have applied to do so. In anticipation of the new Order, most other schools began to adjust their existing provision. Given the delay in implementing the new Order for technology, SCAA recommend that those working to the existing Order should concentrate on Te2 and Te3. The new Order was finally issued at the end of 1994.

91   At a time of such considerable change, inspectors need to be sympathetic to schools' efforts to interpret the new Order and to modify their curriculum in design and technology in preparation for implementation in September 1995.

GCSE

92   The requirements for pupils starting Key Stage 4 in 1994/95 and 1995/96 to study technology are suspended. The National Curriculum requirements remain in force for those pupils who started Y10 in 1993 and schools are therefore under a duty to teach the existing National Curriculum Programme of Study to that cohort of pupils. It is not, though, a statutory requirement that these pupils be assessed in technology, but it is necessary for schools to report on their progress and achievement.

93   Schools are responding in different ways to this interregnum. A small number are transferring pupils, who started National Curriculum design and technology, onto pre-National Curriculum courses, especially where they are also taking a separate short course in IT or a full course in information systems. In a substantial proportion of schools, pupils in this year's Year 10 will be taking pre-National Curriculum courses such as craft, design and technology and home

economics, and in some cases, as in the past, it will be optional rather than mandatory.

GNVQ

94   Although numbers are still small, an increasing number of schools are becoming involved in design and technology-related GNVQ courses. Design and technology teachers may well be involved in a variety of GNVQ programmes such as manufacturing, engineering, and health and social care. The structure of GNVQ courses and the associated assessment requirements are significantly different from those for GCSE/GCE and inspectors will need to be familiar with these differences.

Annex

GCSE results for 15 year olds[1] for Design and Technology 1994

Type of School		Number of 15 year old pupils entered	Percentages achieving grades									Average points score[3]	% A*–C grades	% A*–G grades	Average points score[3]	% A–C grades	Average points score[3]	% A–C grades
			A*	A	B	C	D	E	F	G	U	**1994**			**1993**		**1992[2]**	
Comprehensive		157645	1.5	6.1	12.2	17.0	18.3	16.8	12.7	6.8	2.4	4.03	36.8	91.5	3.67	36.2	#	#
Selective		4182	5.5	18.7	25.9	23.5	13.3	7.1	3.2	1.3	0.2	5.38	73.5	98.4	5.24	72.4	#	#
Modern		5129	0.4	2.0	8.3	15.6	19.3	19.8	15.4	8.3	3.1	3.60	26.3	89.1	3.33	28.0	#	#
Maintained	All pupils	166956	1.5	6.3	12.4	17.2	18.2	16.7	12.6	6.7	2.4	4.05	37.4	91.6	3.69	36.8	#	#
	Boys	128166	1.2	5.2	11.0	16.3	18.5	17.8	13.6	7.4	2.6	3.91	33.6	90.9	3.57	33.9	#	#
	Girls	38790	2.6	10.0	17.2	20.0	17.4	13.0	9.0	4.7	1.6	4.53	49.9	94.1	4.14	47.1	#	#
All Subjects Maintained	All pupils		2.1	8.4	16.4	20.5	18.9	14.5	10.2	4.5	1.5	4.40	47.4	95.5	4.12	46.3	4.14	45.0

1 Aged 15 on 31/8/93
2 1992 results include a small amount of data from special schools
3 Calculated on basis A*=8, A=7, B=6, C=5, D=4, E=3, F=2, G=1

– less than 100 candidates
* more than 100 and less than 500 candidates
x information not available

GCSE results for 15 year olds[1] for Home Economics 1994

Type of School		Number of 15 year old pupils entered	Percentages achieving grades									Average points score[3]	% A*-C grades	% A*-G grades	Average points score[3]	% A-C grades	Average points score[3]	% A-C grades
			A*	A	B	C	D	E	F	G	U	1994	1994	1994	1993	1993	1992[2]	1992[2]
Comprehensive		86468	0.8	5.5	12.5	18.6	20.6	18.3	12.3	5.4	2.0	4.04	37.3	93.9	3.76	36.5	3.86	37.4
Selective		1857	7.4	25.4	32.9	20.0	8.3	1.5	0.3	0.0		5.86	85.7	99.2	5.74	85.1	5.75	85.5
Modern		3773	0.7	2.7	7.6	17.1	23.0	23.4	15.3	5.1	1.7	3.72	28.0	94.8	3.53	28.8	3.52	26.8
Maintained	All pupils	92098	0.9	5.8	12.7	18.5	20.5	18.2	12.2	5.3	1.9	4.07	37.9	94.1	3.79	37.2	3.87	37.6
	Boys	12704	0.3	2.1	6.1	12.3	19.4	22.9	19.3	9.4	3.3	3.37	20.7	91.8	3.16	22.0	3.23	22.1
	Girls	79394	1.0	6.3	13.8	19.5	20.6	17.4	11.1	4.6	1.7	4.18	40.7	94.4	3.88	39.5	3.97	39.9
All Subjects Maintained	All pupils		2.1	8.4	16.4	20.5	18.9	14.5	10.2	4.5	1.5	4.40	47.4	95.5	4.12	46.3	4.14	45.0

1 Aged 15 on 31/8/93

2 1992 results include a small amount of data from special schools

3 Calculated on basis A*=8, A=7, B=6, C=5, D=4, E=3, F=2, G=1

− less than 100 candidates

* more than 100 and less than 500 candidates

x information not available

GCSE results for 15 year olds[1] for other workshop subjects 1994

Type of School		Number of 15[1] year old pupils entered	1994 Percentages achieving grades									Average points score[3]	% A*-C grades	% A*-G grades	1993 Average points score[3]	1993 % A-C grades	1992[2] Average points score[3]	1992[2] % A-C grades
			A*	A	B	C	D	E	F	G	U							
Comprehensive		4032	0.5	2.2	5.2	11.8	20.2	22.3	16.9	8.3	3.7	3.43	19.7	87.4	3.01	24.2	#	#
Selective		69	–	–	–	–	–	–	–	–	–	–	–	–	–	–	#	#
Modern		225	0.4	0.4	4.0	11.6	20.4	24.9	17.3	8.0	0.4	3.30	16.4	87.1	3.01	25.3	#	#
Maintained	All pupils	4326	0.5	2.2	5.4	11.9	20.1	22.3	16.8	8.2	3.5	3.44	20.0	87.4	3.01	24.3	#	#
	Boys	3964	0.6	2.3	5.1	11.3	20.1	22.3	17.3	8.2	3.7	3.42	19.3	87.3	3.00	24.1	#	#
	Girls	362	0.3	1.1	7.7	18.5	20.2	22.4	11.3	7.5	1.4	3.67	27.6	89.0	3.15	26.5	#	#
All Subjects Maintained	All pupils		2.1	8.4	16.4	20.5	18.9	14.5	10.2	4.5	1.5	4.40	47.4	95.5	4.12	46.3	4.14	45.0

1 Aged 15 on 31/8/93
2 1992 results include a small amount of data from special schools
3 Calculated on basis A*=8, A=7, B=6, C=5, D=4, E=3, F=2, G=1

– less than 100 candidates
* more than 100 and less than 500 candidates
x information not available

GCSE results for 15 year olds¹ for Vocational Studies 1994

Type of School		Number of 15¹ year old pupils entered	1994 Percentages achieving grades									Average points score³	% A*–C grades	% A*–G grades	1993 Average points score³	% A–C grades	1992² Average points score³	% A–C grades
			A*	A	B	C	D	E	F	G	U							
Comprehensive		35711	1.4	8.4	14.6	21.0	18.2	15.0	10.7	5.4	2.2	4.30	45.4	94.7	4.02	42.8	3.88	39.0
Selective		101	6.9	36.6	33.7	11.9	7.9	3.0	0.0	0.0	0.0	6.14	89.1	100.0	5.95	89.1	5.00	66.1
Modern		2042	0.3	4.5	10.1	23.1	22.1	17.3	12.1	5.1	2.2	4.01	38.0	94.6	3.78	34.8	3.82	34.0
Maintained	All pupils	37854	1.3	8.3	14.4	21.1	18.4	15.1	10.7	5.4	2.2	4.29	45.1	94.7	4.01	42.4	3.89	38.8
	Boys	11706	0.9	5.6	11.7	18.1	17.9	16.3	13.9	8.1	3.5	3.93	36.2	92.5	3.62	34.6	3.51	31.0
	Girls	26148	1.5	9.5	15.5	22.5	18.6	14.5	9.3	4.2	1.6	4.44	49.0	95.7	4.17	45.7	4.04	41.9
All Subjects Maintained	All pupils		2.1	8.4	16.4	20.5	18.9	14.5	10.2	4.5	1.5	4.40	47.4	95.5	4.12	46.3	4.14	45.0

1 Aged 15 on 31/8/93
2 1992 results include a small amount of data from special schools
3 Calculated on basis A*=8, A=7, B=6, C=5, D=4, E=3, F=2, G=1

− less than 100 candidates
* more than 100 and less than 500 candidates
x information not available

GCSE results for 15 year olds[1] for Design & Technology, Home Economics, other workshop subjects and Vocational Studies 1994

1994

Type of School		Number of 15¹ year old pupils entered	A*	A	B	Percentages achieving grades C	D	E	F	G	U	Average points score[3]	% A*–C grades	% A*–G grades
Comprehensive		283856	6.1	12.5	17.9	19.0	17.1	12.4	6.2	2.3	5.1	4.91	55.6	93.7
Selective		6209	20.8	28.0	22.2	11.7	6.0	2.8	1.0	0.1	1.3	6.36	82.8	92.6
Modern		11169	2.7	8.3	17.4	21.1	20.6	14.8	6.6	2.4	5.5	4.60	49.4	94.0
Maintained	All pupils	301234	6.3	12.7	18.0	19.0	17.0	12.3	6.2	2.2	5.0	4.93	55.9	93.6
	Boys	156540	4.9	10.5	16.0	18.6	18.2	14.2	7.6	2.8	6.2	4.69	49.9	92.7
	Girls	144694	1.6	7.9	15.0	20.2	19.4	15.7	10.2	4.6	1.7	4.32	44.6	94.5
All Subjects Maintained	All pupils		2.1	8.4	16.4	20.5	18.9	14.5	10.2	4.5	1.5	4.40	47.4	95.5

1 Aged 15 on 31/8/93

2 1992 results include a small amount of data from special schools

3 Calculated on basis A*=8, A=7, B=6, C=5, D=4, E=3, F=2, G=1

– less than 100 candidates

* more than 100 and less than 500 candidates

x information not available

GCE AS results for Design and Technology 1994

Type of School		Number of candidates	1994							% A–B grades	% A–E grades	Average points score p	1993		1992	
			Percentages achieving grades										% A–B grades	% A–E grades	% A–B grades	% A–E grades
			A	B	C	D	E	N	U							
Maintained	All pupils	202	9.9	11.4	19.3	21.8	15.8	8.4	9.4	21.3	78.2	2.1	14.5	67.5	#	#
	Boys	152	6.6	10.5	19.7	22.4	17.1	9.9	9.9	17.1	76.3	2.0	13.1	67.7	#	#
	Girls	50	–	–	–	–	–	–	–	–	–	–	–	–	#	#
All subjects Maintained	All pupils		7.1	10.2	14.8	17.9	18.2	12.9	15.1	17.3	68.2	1.8	17.0	65.5	16.6	65.4

– less than 100 candidates

* more than 100 and less than 500 candidates

p Calculated on basis A=5, B=4, C=3, D=2, E=1

The number of pupils taking AS levels is insufficient to yield a meaningful analysis by type of maintained school

GCE AS results for Home Economics 1994

Type of School		Number of candidates	1994 Percentages achieving grades							% A–B grades	% A–E grades	Average points score[p]	1993 % A–B grades	1993 % A–E grades	1992 % A–B grades	1992 % A–E grades
			A	B	C	D	E	N	U							
Maintained	All pupils	144	9.0	6.3	9.7	15.3	18.1	16.7	24.3	15.3	58.3	1.5	14.1	51.3	22.5	61.3
	Boys	15	–	–	–	–	–	–	–	–	–	–	–	–	–	–
	Girls	129	9.3	5.4	10.1	17.1	18.6	17.1	21.7	14.7	60.5	1.5	13.9	53.3	22.5	61.9
All subjects Maintained	All pupils		7.1	10.2	14.8	17.9	18.2	12.9	15.1	17.3	68.2	1.8	17.0	65.5	16.6	65.4

– less than 100 candidates

* more than 100 and less than 500 candidates

p Calculated on basis A=5, B=4, C=3, D=2, E=1

The number of pupils taking AS levels is insufficient to yield a meaningful analysis by type of maintained school

GCE AS results for Vocational Studies* 1994

Type of School		Number of candidates	1994 Percentages achieving grades							% A–B grades	% A–E grades	Average points score[p]	1993 % A–B grades	1993 % A–E grades	1992 % A–B grades	1992 % A–E grades
			A	B	C	D	E	N	U							
Maintained	All pupils	232	5.6	9.1	12.5	22.0	18.5	12.9	15.9	14.7	67.7	1.6	12.7	63.3	18.5	74.2
	Boys	95	–	–	–	–	–	–	–	–	–	–	9.6	55.8	22.9	73.4
	Girls	137	7.3	11.7	15.3	21.9	15.3	12.4	12.4	19.0	71.5	1.9	15.0	69.2	15.7	74.7
All subjects																
Maintained	All pupils		7.1	10.2	14.8	17.9	18.2	12.9	15.1	17.3	68.2	1.8	17.0	65.5	16.6	65.4

– less than 100 candidates

* more than 100 and less than 500 candidates

p Calculated on basis A=5, B=4, C=3, D=2, E=1

The number of pupils taking AS levels is insufficient to yield a meaningful analysis by type of maintained school

GCE AS results for Design & Technology, Home Economics, other workshop subjects and Vocational Studies 1994

Type of School		Number of candidates	Percentages achieving grades							% A–B grades	% A–E grades	Average points score[p]
			A	B	C	D	E	N	U			
Maintained	All pupils	578	8.0	9.2	14.2	20.2	17.5	12.3	15.7	17.1	69.0	1.8
	Boys	262	5.3	8.8	14.9	21.0	19.1	11.5	16.0	14.1	69.1	1.7
	Girls	316	10.1	9.5	13.6	19.6	16.1	13.0	15.5	19.6	69.0	1.9
All subjects												
Maintained	All pupils		7.1	10.2	14.8	17.9	18.2	12.9	15.1	17.3	68.2	1.8

– less than 100 candidates

* more than 100 and less than 500 candidates

p Calculated on basis A=5, B=4, C=3, D=2, E=1

The number of pupils taking AS levels is insufficient to yield a meaningful analysis by type of maintained school

42

GCE A-Level results for Design & Technology 1994

Type of School		Number of candidates	Percentages achieving grades							% A–B grades	% A–E grades	1993 % A–B grades	1993 % A–E grades	1992 % A–B grades	1992 % A–E grades
			A	B	C	D	E	N	U						
Comprehensive		5196	8.5	12.9	19.7	22.9	17.3	9.7	6.5	21.4	81.4	21.5	80.8	#	#
Selective		517	13.7	16.6	26.9	18.4	13.2	5.4	4.4	30.4	88.8	28.8	84.5	#	#
Modern		47	–	–	–	–	–	–	–	–	–	–	–	#	#
Maintained	All pupils	5760	8.9	13.2	20.4	22.4	17.0	9.4	6.4	22.1	81.9	22.2	81.1	#	#
	Boys	4643	8.7	12.7	20.1	22.2	17.4	9.7	6.7	21.4	81.2	21.8	80.3	#	#
	Girls	1117	10.0	15.1	21.6	22.8	15.2	8.2	4.9	25.2	84.8	23.9	84.6	#	#
All subjects															
Maintained	All pupils		13.1	16.2	18.5	18.9	15.2	9.4	7.5	29.3	81.9	14.5	67.5		#

– less than 100 candidates

* more than 100 and less than 500 candidates

GCE A-Level results for Home Economics 1994

Type of School		Number of candidates	1994 Percentages achieving grades							% A–B grades	% A–E grades	1993 % A–B grades	% A–E grades	1992 % A–B grades	% A–E grades
			A	B	C	D	E	N	U						
Comprehensive		1412	8.4	12.6	19.2	23.4	16.8	10.3	8.4	21.0	80.4	19.8	79.6	19.4	75.9
Selective		244	21.3	14.3	18.4	18.9	7.8	11.1	7.8	35.7	80.7	30.7	86.5	31.3	87.7
Modern		6	–	–	–	–	–	–	–	–	–	–	–	–	–
Maintained	All pupils	1662	10.3	12.9	19.1	22.6	15.5	10.4	8.3	23.2	80.4	21.4	80.5	21.0	77.7
	Boys	85	–	–	–	–	–	–	–	–	–	–	–	–	–
	Girls	1577	10.7	12.8	19.3	23.0	15.2	10.3	7.9	23.5	80.9	21.6	81.2	21.1	78.1
All subjects Maintained	All pupils		13.1	16.2	18.5	18.9	15.2	9.4	7.5	29.3	81.9	28.0	79.7	26.4	78.6

– less than 100 candidates

* more than 100 and less than 500 candidates

GCE A-Level results for Vocational Studies 1994

Type of School		Number of candidates	1994 Percentages achieving grades A	B	C	D	E	N	U	% A–B grades	% A–E grades	1993 % A–B grades	1993 % A–E grades	1992 % A–B grades	1992 % A–E grades
Comprehensive		821	6.6	15.2	17.4	16.1	16.7	11.7	14.7	21.8	72.0	22.2	70.3	22.1	74.0
Selective		48	–	–	–	–	–	–	–	–	–	–	–	–	–
Modern		8	–	–	–	–	–	–	–	–	–	–	–	21.8	68.2
Maintained	All pupils	877	7.3	15.6	17.3	16.2	16.5	11.5	14.0	22.9	73.0	23.2	71.4	23.2	73.9
	Boys	394	6.3	15.2	15.2	17.3	16.0	11.9	17.0	21.6	70.1	20.8	66.0	22.4	69.7
	Girls	483	8.1	15.9	19.0	15.3	17.0	11.2	11.6	24.0	75.4	24.7	74.8	23.7	76.8
All subjects Maintained	All pupils		13.1	16.2	18.5	18.9	15.2	9.4	7.5	29.3	81.9	28.0	79.7	26.4	78.6

– less than 100 candidates

* more than 100 and less than 500 candidates

GCE A-Level results for Design & Technology, Home Economics, other workshop subjects and Vocational Studies 1994

Type of School		Number of candidates			Percentages achieving grades						% A–B grades	% A–E grades
			A	B	C	D	E	N	U			
Comprehensive		7429	8.3	13.1	19.4	22.2	17.2	10.1	7.8	21.4	80.1
Selective		809	16.4	16.3	23.7	18.2	11.7	7.3	5.3	32.8	86.4
Modern		61	–	–	–	–	–	–	–	–	–
Maintained	All pupils	8299	9.0	13.4	19.8	21.8	16.7	9.8	7.6	22.4	80.6
	Boys	5122	8.4	12.9	19.7	21.8	17.4	9.9	7.7	21.3	80.1
	Girls	3177	10.1	14.1	20.1	21.8	15.5	9.7	7.4	24.2	81.6
All subjects												
Maintained	All pupils		13.1	16.2	18.5	18.9	15.2	9.4	7.5	29.3	81.9

– less than 100 candidates

* more than 100 and less than 500 candidates